ALEKS User's Guide

With Two Term Access Code

Harold D. Baker, Ph.D.

ALEKS Corporation

Boston   Burr Ridge, IL   Dubuque, IA   Madison, WI   New York   San Francisco   St. Louis
Bangkok   Bogotá   Caracas   Lisbon   London   Madrid
Mexico City   Milan   New Delhi   Seoul   Singapore   Sydney   Taipei   Toronto

McGraw-Hill Higher Education

A Division of The McGraw-Hill Companies

ALEKS USER'S GUIDE

ALEKS is a trademark of ALEKS Corporation in the United States and other countries.

Copyright ©2000 by ALEKS Corporation. All rights reserved. Printed in the United States of America. Except as permitted under the United States Copyright Act of 1976, no part of this publication may be reproduced or distributed in any form or by any means, or stored in a data base or retrieval system, without the prior written permission of ALEKS Corporation.

ALEKS Corporation may have patents, patent applications, trademarks, copyrights, or other intellectual property rights covering subject matter in this Manual. Except as expressly provided in any written agreement from ALEKS Corporation, the furnishing of this Manual does not give you any license or other right to these patents, trademarks, copyrights, or other intellectual property.

Information in this Manual is subject to change without notice.

The names of companies and products mentioned in this Manual may be the trademarks of their respective owners.

Netscape® and Netscape Communicator are trademarks of Netscape Communications Corporation.

This book is printed on acid-free paper.

9 0 COU COU 0

ISBN-13: 978-0-07-243548-1
ISBN-10: 0-07-243548-8

Vice president and editorial director: *Kevin T. Kane*
Publisher: *JP Lenney*
Sponsoring editor: *William K. Barter*
Developmental editor: *Byron D. Smith*
Marketing manager: *Mary K. Kittell*
Media technology project manager: *Steve Metz*
Project manager: *Renee C. Russian*
Production supervisor: *Sandy Ludovissy*
Designer: *K. Wayne Harms*
Supplement coordinator: *Tammy Juran*
Printer: *Courier*

Cover designers: *John and Kim Rokusek*

www.mhhe.com

Contents

1	**Preface**	1
2	**Technical Requirements**	1
3	**Registration & Installation**	2
4	**Tutorial**	5
5	**Assessments and Learning**	5
	5.1 Assessments	5
	5.2 Results	7
	5.3 Learning Mode	8
	5.4 Progress in the Learning Mode	8
	5.5 Additional Features	8
6	**Logging on to Your Account**	9
7	**Installation on Additional Machines**	10
8	**Guidelines for Effective Use**	10
9	**Frequently Asked Questions**	11
10	**Troubleshooting**	17

1 Preface

Welcome to **ALEKS**! You are about to discover one of the most powerful educational tools available for college mathematics. Combining advanced learning technology with the flexibility of the World Wide Web, the **ALEKS** system provides a "smart" interactive tutoring system with unmatched features and capabilities. Richly supplied with illustrations and reference materials, **ALEKS** constantly challenges you and supplies extensive feedback on what you have accomplished. **ALEKS** will always help you select the ideal topic to work on now. That way you learn concepts in the order that's best for you. **ALEKS** provides individualized, one-on-one instruction that fits your schedule. It is available wherever you access the Web.

ALEKS was developed with support from the National Science Foundation. It is based on a field of Mathematical Cognitive Science called "Knowledge Spaces." The purpose of research in Knowledge Spaces is to model human knowledge of any subject for quick and precise assessment by interactive computer programs.

The **ALEKS** system is self-explanatory and includes online instructions and feedback. This booklet contains basic information to help students begin using **ALEKS**. Instructors using **ALEKS** with their courses are provided with an *Instructor's Manual* containing complete information on the system's operation. They should be able to answer any questions beyond those dealt with in these pages.

NOTE: ALEKS is designed for use without help from a manual. Your instructor will assist you in registering with the system and beginning to use it. If questions arise, or if you want to learn more about ALEKS, use this *Guide*. It is intended as a convenient and concise reference.

Only registered users can keep an account on **ALEKS**. (Anyone may try the system as a guest.) Two or more persons cannot use the same ALEKS account. The system will regard them as a single person and give incorrect guidance.

2 Technical Requirements

PC Requirements

You can use **ALEKS** on any PC with a Pentium processor of 133 MHz or more (166+ MHz preferred) or any Pentium II or Pentium III processor. At least 32 MB of RAM are required. Your operating system must be Windows 95/98/NT4.0 or higher.

The following popular web browsers are compatible with **ALEKS** on PCs: Netscape Communicator 4.5 or higher, Internet Explorer 4.0 or higher.

Macintosh Requirements

ALEKS can be used on a Macintosh, iMac, or Macintosh G3-G4 with at least 32 MB of RAM. Your operating system must be MacOS 7.6.1 or higher.

Only Netscape Communicator 4.5 and higher is supported for the Macintosh.

Note

Netscape Communicator 4.0x (4.01, 4.02, etc.) is not compatible with **ALEKS** and should be upgraded to 4.5 or higher. This can be done from the Netscape website:

> http://www.netscape.com

Internet Access

ALEKS is used over the World Wide Web. You must have an Internet connection by dialup modem (at least 28k) or any other kind of access to the Internet (cable, ISDN, DSL, etc.).

America Online Subscribers

If you have America Online 3.0 you will have to upgrade to America Online 4.0 or higher in order to use **ALEKS**. You can upgrade from AOL.

3 Registration & Installation

Before You Begin: In order to register as an **ALEKS** user you need the **Access Code** inside the back cover of this booklet. You also need a **Course Code** provided by your instructor. When you register with the **ALEKS** system your name is entered into the database and records of your progress are kept. If the **ALEKS** plugin has not been installed on the computer being used for registration, it will be installed automatically as part of this procedure.

Step 1: Go to the **ALEKS** website for Higher Education by typing in the following address:

> http://www.highed.aleks.com

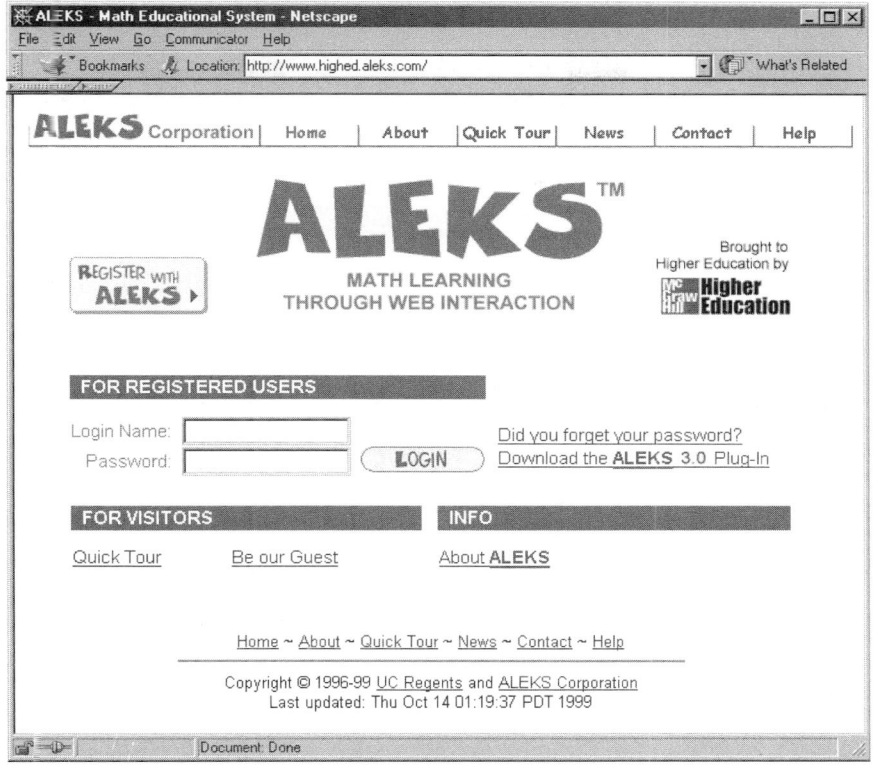

Figure 1: The ALEKS Website for Higher Education

NOTE: If you are typing this URL by hand, pay careful attention to the spelling "aleks." Also, the other **ALEKS** websites you might find using a search engine will not work for you. You will be able to register *only* at the address given above.

For your convenience, add a "Bookmark" or "Favorite" at this location. This is the site where you will log in to your account.

Step 2: Click on "Register with ALEKS" (Fig. 1).

Step 3: You will see instructions for students and instructors registering with **ALEKS**. Click on "Register" where it says "For Students" (on the left-hand side).

NOTE: If you do not have a current plugin, one will be installed. Do not interrupt this process until a message appears saying that the installation is complete. Then you will need to quit your Web browser ("Exit,"

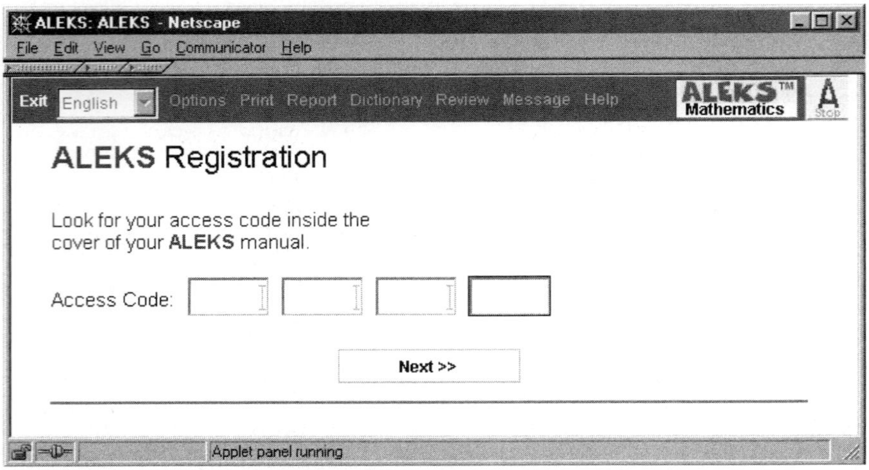

Figure 2: Access Code

"Close," or "Quit" under the "File" menu) and open your Web browser again. Then go back to the **ALEKS** website for Higher Education (use your Bookmark/Favorite). Return to Step 1, above, to begin registration.

Step 4: At the beginning of registration you will be asked for your **Access Code**. It is on a sticker inside the back cover of this booklet. Enter the Access Code in the spaces provided and click on "Next" (Fig. 2). Answer the questions to complete your registration. Among other questions, you will be asked to enter your email address. Supplying this information enables your site administrator to help you with problems more quickly. You will also be asked to enter your Student ID number. Please double-check this number before clicking on "Next."

Step 5: At the end of registration you will be given a Login Name and Password. Write these down and keep them in a safe place. You will need them to return to the system (Sec. 6). Your Login Name is not the same as your name. It usually consists of the first letter of your first name plus your whole last name run together, with no spaces or punctuation. Thus "Jane Smith" may have the Login Name "jsmith"; if there is more than one "Jane Smith" in the database, a numeral will be appended, as "jsmith2." You can change your Password at any time (Sec. 5.5).

NOTE: Your Login Name and Password can be typed with upper- or lower-case letters. Neither may contain spaces or punctuation.

Step 6: Following registration you will be asked for your **Course Code**.

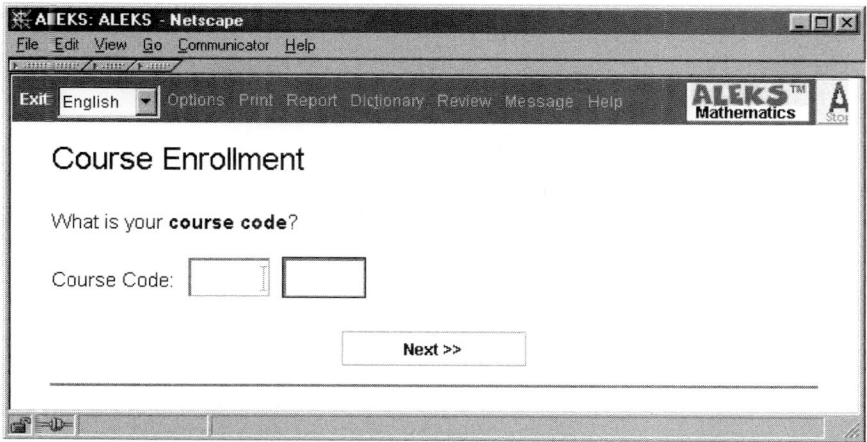

Figure 3: Course Code

The Course Code is supplied by your instructor. Enter this in the spaces provided and click on "Next" (Fig. 3). Now you can begin the Tutorial.

4 Tutorial

The **ALEKS** system does not use multiple-choice questions. All answers are mathematical expressions and constructions. After registration, the **ALEKS** Tutorial will teach you to use the simple tools needed for your course (Fig. 4). There is plenty of feedback to help you complete it successfully.

NOTE: The Tutorial is not intended to teach mathematics. It just trains you to use the **ALEKS** input tool (called the "Answer Editor"). The correct input is always shown, and you simply enter what you see. Online help is also available while you are using **ALEKS** by clicking the "Help" button (Sec. 5.5).

5 Assessments and Learning

5.1 Assessments

Instruction through **ALEKS** is guided by precise understanding of your knowledge of the subject. This information is obtained by assessments in

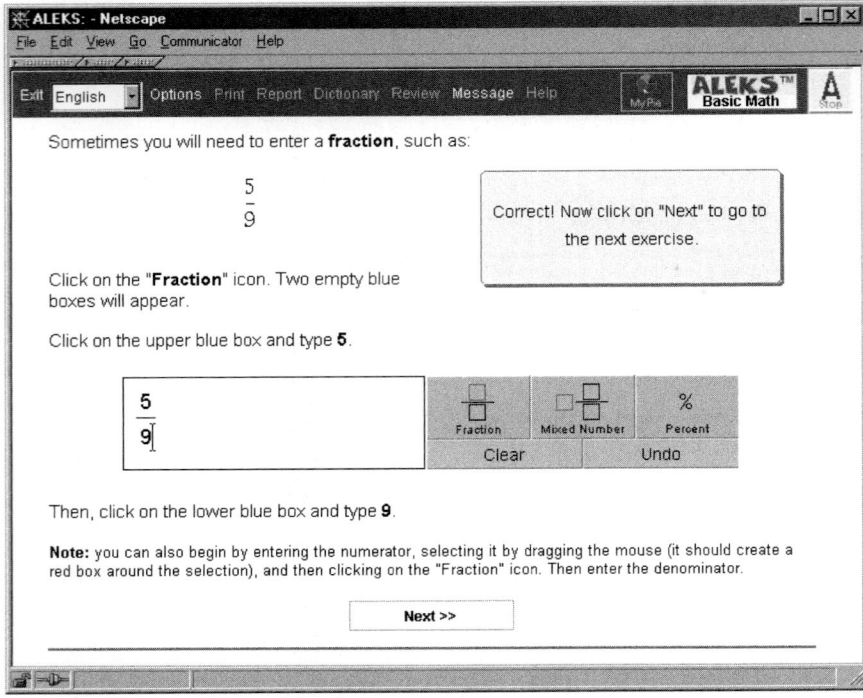

Figure 4: The Answer Editor (Tutorial)

which the system asks you to solve a series of problems. (The system's estimate of your knowledge is also updated when you make progress in the Learning Mode.) Your first assessment occurs immediately after the Registration and Tutorial. If you are enrolled in a course covering more than one subject you may go through multiple assessments. There will be one assessment for each subject.

NOTE: Your instructor may require that the first assessment be taken under supervision. **Don't try begin your initial assessment at home until you find out where your instructor wants you to take it.** Additional assessments may be scheduled for you by the instructor. These may or may not need to be supervised, depending on the instructor's preference. The **ALEKS** system also prompts "automatic" assessments when you have spent a certain amount of time on the system or have made a certain amount of progress.

Before the start of your first assessment you will be asked to estimate how well you know the subject. This information may make your assessment

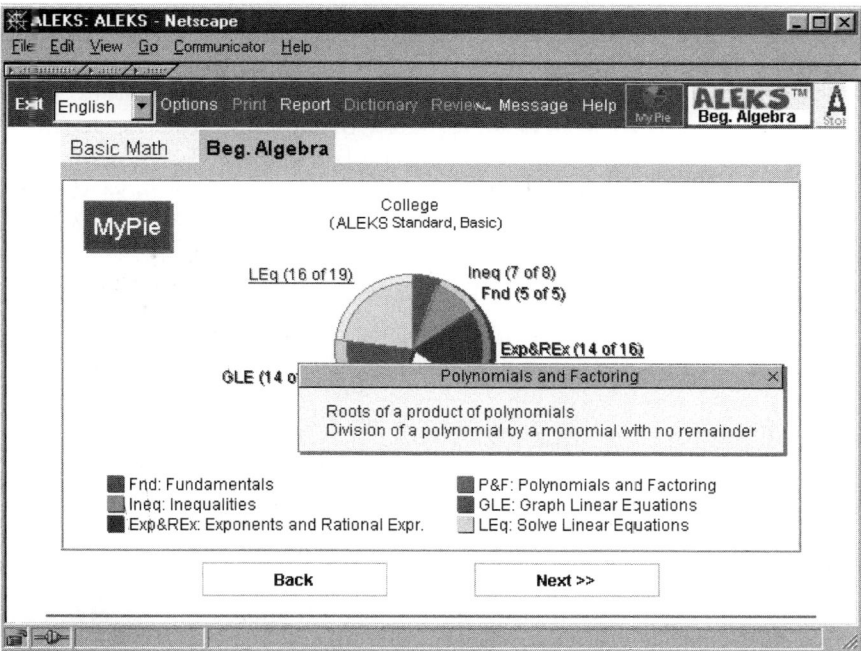

Figure 5: Assessment Report

shorter, but it has no effect on the outcome or on the level where you will begin to use **ALEKS**. If you don't know, just select "Unknown."

5.2 Results

Assessment results are presented in the form of color-coded piecharts. Slices of the piecharts correspond to parts of the syllabus. The relative size of the slices represents the importance of each topic for the syllabus. The solidly colored part of a slice indicates how close you are to mastering that part of the syllabus.

NOTE: You may see more than one piechart displayed following an assessment when you are progressing through a series of courses or units. (Your knowledge in the previous and/or subsequent units is also displayed.)

5.3 Learning Mode

Following the presentation of assessment results, the system will display a combined piechart ("MyPie"). This piechart shows the entire syllabus through the end of your current course. By placing the mouse pointer over slices of the pie, you can see which concepts you are now most ready to learn (Fig. 5). Not all slices will contain concepts at any given time. They may have been mastered already, or work may need to be done in other slices before they become available. The concept you click on becomes your entry into the Learning Mode. The system will help you in seeking to master that concept and "add it to your pie."

5.4 Progress in the Learning Mode

In the Learning Mode, you are given practice problems based on the chosen topic. You also get explanations of how to solve this kind of problem and access to a dictionary of mathematical concepts. Underlined mathematical terms are links to the dictionary. Click on any term to get a complete definition. The system will require a number of correct answers before it assumes that you have mastered the concept. Then it "adds it to your pie." At this point a revised piechart will be shown reflecting your new knowledge. You will be able to choose a new concept to begin. If you make mistakes, more correct answers may be required. If you tire of this topic and wish to choose another, click on "MyPie" near the top of the window. This will make you exit the topic and you will get the piechart for a new choice. If you make repeated errors on a given concept, the system will conclude that the concept was not mastered. It will offer you a new choice of more basic concepts.

NOTE: Let **ALEKS** do its job! It is normal to have trouble mastering new concepts the first time around. When this happens, the system responds by revising its view of your knowledge and offers new choices. Keep in mind that the system does not "drill" you on concepts it believes you already know. The concepts presented as most "ready to learn" are always those just at the edge of your current knowledge. These are the topics you are completely prepared to learn.

5.5 Additional Features

All buttons described below are available in the Learning Mode. In the Assessment Mode, only the "Exit" and "Help" buttons are active.

Changing Your Password

If you want to change your Password, click on the "Options" button. Here you can also choose to look at your last Assessment Report and the work you've done recently in Learning Mode.

Report

Any time you wish to look at your assessment reports, click on "Report." Choose any date from the menu and click "Graph."

Dictionary

To search the online dictionary of mathematical terms, click "Dictionary."

Review

To review past material, click on the "Review" button. You will be offered a choice of concepts to practice. An automatic review will also be prompted, if needed, when you log on.

Messages

Your instructor can send you messages via **ALEKS**. You see new messages when you log on. You can also check for messages by clicking on the "Message" button. Your instructor can choose to let you reply to messages as well.

Help

For online help with the use of the Answer Editor, click "Help."

MyPie

Clicking "MyPie" gives you a piechart summarizing your current mastery. You can use this piechart to choose a new concept.

6 Logging on to Your Account

Step 1: You always log on from the **ALEKS** website for Higher Education:

> http://www.highed.aleks.com

Use the "Bookmark" or "Favorite" for this site if you made one (Sec. 3). Remember that you may find other **ALEKS** websites via a search engine, but this is the only one with your account.

Step 2: On the login page enter the Login Name and Password provided at the time of registration (Sec. 3, Step 5). Be sure to type these correctly, without any spaces or punctuation.

Step 3: If you enter your Login Name and Password correctly, your browser will begin accessing the plugin to start **ALEKS**. This takes a few seconds. You will then come to the place you left off in your previous **ALEKS** session.

NOTE: If you forget your Login Name or Password contact your instructor. It is a good idea to change your Password to one you will remember easily but is difficult for others to guess (Sec. 5.5).

7 Installation on Additional Machines

Before You Begin: Installing **ALEKS** means installing the **ALEKS** plugin. This is the software used by your web browser to access and run **ALEKS**. You can access your **ALEKS** account from any computer that meets the technical requirements and has had the **ALEKS** plugin installed. You cannot use **ALEKS** without the **ALEKS** plugin that is installed over the World Wide Web.

Step 1: Go to the **ALEKS** website for Higher Education:

>   http://www.highed.aleks.com

Add a "Bookmark" or "Favorite" at this location.

Step 2: Use your Login Name and Password to log in (Sec. 3, Step 5).

Step 3: When you log on to **ALEKS**, the system will automatically check to see if your system is compatible and if you have the most recent version of the **ALEKS** plugin. If you do not have a current plugin, it will download the plugin and ask your permission to install. After you grant permission, it will install the (new) plugin. Do not interrupt the installation process until a message appears stating that the installation is complete and asks you to restart your browser. You will need to quit your Web browser ("Exit," "Close," or "Quit" under the "File" menu), open your Web browser again, and go back to the **ALEKS** website for Higher Education (use your Bookmark/Favorite).

8 Guidelines for Effective Use

Supplementary Materials

You should have pencil and paper ready for all assessments and use in the Learning Mode. Basic calculators should be used only when you are instructed to do so. (A basic calculator is part of **ALEKS**.)

Assessments

You should not ask for, nor receive any help during assessments. Not even explanations or rephrasing of problems are permitted. If you receive help, the system will get a wrong idea of what you are most ready to learn, and this will hold up your progress. Any time you are unsure of something, click "I don't know." (Don't guess!)

Learning Mode

You should learn to use the special features of the Learning Mode, especially the explanations and the mathematical dictionary. A button marked "Ask a Friend" may also appear from time to time. Clicking on this button will prompt the system to suggest the name of a classmate who has mastered the concept.

Regular Use

Nothing is more important to your progress than regular use of the system. Three hours per week is a recommended *minimum*. Five is better. Put **ALEKS** into your weekly schedule and stick to it!

9 Frequently Asked Questions

What are the rules for taking an assessment in ALEKS?

[**Sec. 8**] You must have paper and pencil when taking an assessment in **ALEKS**. For Basic Math, no calculator is permitted. For Algebra, you should also have a simple calculator (no graphing or symbolic functions). A basic calculator is part of **ALEKS**. No help whatsoever is permitted, not even to the extent of rephrasing a problem. Cheating is not a danger, since students are given different problem-types in different sequences. Even if, by chance, two students sitting next to one another were to get the same problem-type at the same time, the actual problems would almost certainly have different numerical values and require different answers. During the assessment, you are not told if your answer is right or wrong. In the Learning Mode, however, you are always told if you make a mistake, and often what that mistake was. The assessment is not a test. Its main purpose is to determine what you are most ready to learn and help you make the best progress possible toward mastery.

How do I add concepts to my pie?

[**Sec. 5.4**] You fill in your pie and achieve mastery in the subject matter by working in the Learning Mode on concepts and skills that the assessment has determined you are most "ready to learn." When you master a concept in the Learning Mode by successfully solving an appropriate number of problems, you will see that your piechart has been changed by the addition of that concept. The goal is to fill in the pie completely.

Why is it that I mastered all the concepts in the Learning Mode, but my assessment says I still have concepts to learn?

In the Learning Mode you are always working on one concept at a time, whereas assessments are cumulative and "test" you on everything in the given subject matter. It may be more difficult to show mastery of concepts you have recently worked on, when you are being quizzed on many different topics at the same time. For this reason, your assessment results may not exactly match what you had mastered in the Learning Mode. This is normal and simply means that you should keep working in the system. (Sometimes the opposite also occurs. That is, progress in the assessment turns out to be faster than in the Learning Mode.)

Why doesn't my piechart show any concepts from a category if I haven't filled in that category yet?

[**Sec. 5.3**] You are completely "ready to learn" a set of concepts or skills when you have mastered all the prerequisite concepts or skills that they demand. To take an elementary example, in order to learn "addition of two-digit numbers with carry" you might have to first learn "addition of two-digit numbers without carry" and nothing else. Your piechart will not offer you concepts to work on if you are not ideally ready to begin learning them, that is, they have prerequisites you have not yet mastered. For this reason, your piechart may show that you have only mastered 8 out of 10 concepts for a particular slice of the pie (a particular part of the curriculum), but the piechart says you have no concepts available from that slice to work on. This means that the concepts you have left to master have prerequisites in other areas of the curriculum that you must master first. Keep working in the other slices, and eventually the concepts in that slice will become "available."

What is the difference between "Explain" and "Practice"?

When you begin working on a particular concept in the Learning Mode, you will be shown the name of the concept, a sample problem, and a choice between "Practice" and "Explain." If you think you know how to solve the problem, click "Practice." You will be given a chance to solve the same problem that was initially displayed. If you are not sure, click "Explain" to produce an explanation of how to solve the displayed sample problem. At the bottom of the Explanation page you have the "Practice" button, and sometimes other options for more detailed explanations and help. The Explanation page may also contain a link or reference to a McGraw-Hill textbook. If you click the "Practice" button following an explanation, you are offered a different problem of the same type, not the one whose solution was explained. In order to master the concept and add it to your pie, you must successfully solve a certain number of "Practice" problems. If you wish to choose a new concept, click the "MyPie" button on the **ALEKS** menu bar.

How does the Learning Mode help me learn?

[**Sec. 5.4**] In the Learning Mode, do your best to solve the problems that are offered you. Do not lightly change topics or stop before the system tells you that you are done or suggests choosing another concept. Get to know the features of the Learning Mode, especially the explanations and the Dictionary. The Learning Mode will always tell you if your answer is correct or not. In many cases it will provide information on the kind of error you may have made. Pay attention to this feedback and be sure you understand it.

Keep in mind that **ALEKS** is always giving you material that, in its estimation, you are ideally ready to learn. It does not offer material you have already mastered, except in the Review mode. To go back to concepts you have already worked on, click the "Review" button on the **ALEKS** menu bar.

How does ALEKS create practice problems?

ALEKS creates both Assessment and Practice problems by means of computer algorithms, based on the definition of a particular concept or skill to be mastered. Thus, a particular concept or problem-type may serve as the basis for a very large number of specific problems, each with different numerical values and sometimes (as in the case of word problems) differing in other ways as well. With **ALEKS**, you cannot "learn the test" or "teach to the test."

What happens if I don't learn a concept (or get tired of working on a concept)?

[**Sec. 5.4**] You must answer what the system judges to be an appropriate number of Practice problems correctly to add a concept to your pie. If you make mistakes, you must answer more. **ALEKS** will always tell you when you have mastered the concept. You cannot make this decision for yourself. If you wish to stop working on a concept and choose another one, you can click on "MyPie." Keep in mind, however, that when you come back to the former concept you will must start from the beginning with it. It is usually better to do your best to master the concept you are working on, unless the system tells you to switch. If you are clearly not making progress, **ALEKS** will suggest that you choose something else to work on.

Why is ALEKS giving me things we haven't done in the course or that are too hard?

[**Sec. 5.4**] The most common reason that problems seem too difficult is that you received some help in the assessment, and **ALEKS** has an incorrect estimate of your actual knowledge. The problem, however, corrects itself as soon as you stop getting help. When you fail to master several concepts, **ALEKS** will quickly bring you back to a more comfortable place.

Remember that **ALEKS** is designed to give you material that you are ideally prepared to learn. It will not "drill" what has already been mastered, except in the sense that old knowledge is continually being exercised in the acquisition of new knowledge. **ALEKS** has no idea what you have done or are doing in class from one week to the next. In **ALEKS** you follow an individualized path through the curriculum that is produced by your own learning and your own choices.

Why is ALEKS giving me a new assessment?

[**Sec. 5.1**] New assessments are automatically prompted by **ALEKS** when you have spent sufficient time in the Learning Mode or when you have made adequate progress. Your instructor may also request an assessment for you personally, or for everyone in the course. In this case it may be stipulated that the assessment must be taken in the computer lab. (If you attempt to work at home when an assessment has been ordered to be done in the lab, **ALEKS** will tell you that you need to log on from the lab and deny access.)

How can I get a new assessment in ALEKS?

You cannot initiate a new assessment. **ALEKS** or your instructor must make the request.

Why do I need to take a Tutorial to use ALEKS?

[**Sec. 4**] The Tutorial is a brief interactive training program that teaches you to use the **ALEKS** input tools, or "Answer Editor." **ALEKS** does not use multiple-choice questions. Rather, it requires that answers be given in the form of mathematical expressions and geometrical and other constructions. The Answer Editor is a flexible set of tools enabling you to provide such answers. Although the Answer Editor is easy to use, the Tutorial will make sure you are completely proficient with it before beginning the **ALEKS** system. The Tutorial guides you through every step of learning to use the Answer Editor.

What can I do if I make a mistake entering an answer?

If you make an error entering an answer with the Answer Editor, click on "Undo" to go back one step, or on "Clear" to start over. You can also use the "Backspace" key on your keyboard in the usual way.

NOTE: You cannot use "Undo" or the "Back" button on your browser to go back if you have submitted an answer by clicking on "Next." If you realize that the answer you submitted was incorrect, don't be concerned; the system will most likely recognize this as a careless error based on your other answers and make allowances for it.

What are the icon buttons for?

The icon buttons are used to enter mathematical symbols and to create forms for mathematical expressions. In some cases the keyboard equivalents for icon buttons can be used.

Why are the buttons "sticky"?

[**Sec. 10**] The buttons in the **ALEKS** interface may seem "sticky" at first. If so, try clicking them just a bit longer than usual. You will quickly get used to them.

Why doesn't anything appear when I type?

[**Sec. 10**] In order to type input in the Answer Editor you must first click on a blue box. Each blue box in the input area corresponds to a mathematical expression. When you click on an icon button for a complex expression, it may place more than one blue box in the space, one for each part of the expression. Each blue box must be filled in for a complete expression. For instance, when you click on the "Exponent" icon button, you get two blue boxes. The big one is for entering the base, and a smaller one that is raised and to the right is used to enter the exponent.

How do I get help while using ALEKS?

[**Sec. 5.5**] You can get help using the Answer Editor by clicking the "Help" button on the **ALEKS** menu bar.

Can my instructor or friend help me (or can I use a calculator) in the Learning Mode?

[**Sec. 8**] Help and collaboration are allowed in the Learning Mode. Keep in mind, however, that if you get too much help, the system will start giving you problems that you are not prepared to solve. As a general rule, you can get help with one Practice problem, but you should solve the others yourself.

You need paper and pencil for the Learning Mode, just as you did for the assessment. Use of a calculator is permitted in Algebra only (without symbolic or graphing functions). A basic calculator is part of **ALEKS**.

Why are some of the words I see underlined?

[**Sec. 5.5**] Underlined words in the Learning Mode are links to the on-line mathematics dictionary. Click on any underlined word to see its definition. You can also access the Dictionary by clicking the "Dictionary" button on the **ALEKS** menu bar. The Dictionary is not available during assessment.

Note that the Dictionary is opened in a new window. When you are finished reading the definition, close or "Minimize" the window, and you will see the previous screen. Clicking "Back" on the browser won't work.

What is the "Ask a Friend" button for?

[**Sec. 8**] The "Ask a Friend" button sometimes appears when you are having difficulty with a particular concept. When you click on the button, the system suggests the name of a classmate who has mastered the concept and may be able to help you.

How can I change my Password?

[**Sec. 5.5**] You can change your Password by clicking the "Options" button on the **ALEKS** menu bar.

How can I review material I have already worked on?

[**Sec. 5.5**] Click on the "Review" button to work on material you have already spent time on.

How can I see the reports from previous assessments?

[**Sec. 5.5**] To see any of your assessment reports, click on "Report" (on the **ALEKS** menu bar).

How can I choose a new topic to work on?

[**Sec. 5.5**] To see your current piechart and choose a new concept in the Learning Mode, click on "MyPie" (on the **ALEKS** menu bar), move around on the pie, and choose.

How can I print something in ALEKS?

[**Sec. 10**] To print the contents of the screen, click "Print" on the **ALEKS** menu bar. This produces a new, printable window (**ALEKS** output is not normally printable). Depending on your browser, you may also have to click the browser's "Print" button. When you are done, close the new window.

What do I do if it's taking too long for a new page to load (or if the program freezes)?

[**Sec. 10**] It shouldn't take more than a few seconds for **ALEKS** to respond when you click on any button. If you experience delay, freezing, or crashing, your first step is to click on the small "A" button at upper right. If this doesn't work, click your browser's "Reload" or "Refresh" button. If this doesn't work, close your browser and restart it. In extreme cases use Ctrl-Alt-Delete (Cmd-Opt-Esc on Macintosh). You will come back to the exact place you left off after you log back on.

How do I exit the ALEKS program?

To leave **ALEKS**, click the "Exit" button on the **ALEKS** menu bar or simply close your browser. **ALEKS** always remembers where you left off and brings you back to that place.

Why do I have to log on to ALEKS?

[**Sec. 1**] The fact that **ALEKS** is used over the World Wide Web means that you can access it from your college computer lab or from home. As a registered user of **ALEKS**, you have an account on the server that contains a record of all the work you have done. Your instructor and administrators at your college have access to these records. They can monitor your progress and use of the system as well as carry out administrative functions. Web access also means that there is almost no maintenance or technical preparation required—no disks, CDs, peripherals, or backup procedures.

What if I have a question or problem using ALEKS?

If you have a question or problem using **ALEKS** that is not answered here, contact your instructor. Your instructor has been provided with extensive information on the operation of **ALEKS** and should be able to answer almost any question you may have.

What if I forget my Login Name or Password?

If you lose your Login Name or Password, contact your instructor.

10 Troubleshooting

Login Not Successful

First of all, be careful to type your Login Name and Password correctly, with no spaces or punctuation. Then, be sure you have accessed the **ALEKS** website for Higher Education. There is more than one **ALEKS** website, and only the one at which you registered contains your account. Use the URL provided in this booklet rather than looking for "aleks" via a search engine.

Sticky Buttons

Buttons in the **ALEKS** interface respond differently on different platforms. Sometimes you have to hold the mouse button down a bit longer than is usual. With some practice you should become accustomed to it.

Typed Input Does Not Appear

If you have trouble entering numbers or symbols in the Answer Editor, be sure that you have clicked on a blue box and that the pointer is within the answer area (the rectangle containing the blue boxes).

NOTE: It is not always possible to use the number keys on your keyboard's right-hand "keypad" (check that "Num Lock" has been pressed).

Mixed Number Difficulties

The Answer Editor is easy to use. One warning, however: mixed numbers must be entered using the Mixed Number icon, not by entering the whole part and then using the Fraction icon.

Freezing and Slow Response

If you are logged on to **ALEKS** and the program is either not responding or taking too long to load a new page, one of the following three actions may help (try them in the order given):

1. click on the small "A" in the upper right-hand corner of the **ALEKS** window;
2. click on your browser's "Reload" (or "Refresh") button;
3. close the browser and log on again (the system will bring you back to where you left off); if you cannot close the browser use Ctrl-Alt-Delete (PC) or Cmd-Opt-Esc (Macintosh) and end the task (or reboot, if all else fails).

Open applications other than the web browser that you are using to access **ALEKS** are another cause of slowness. Closing these applications may correct the problem.

If slowness persists, it is most likely due to a problem in the local network. Bring this to the attention of your system or lab administrator.

Lengthy Assessment

It is impossible to know how many questions will be asked in an assessment. The number of questions asked does not reflect your knowledge of the subject matter. It may reflect the consistency of your effort or concentration.

Reduction of Piechart

You may observe a loss of concepts in your piechart either in the Learning Mode or following an assessment. This is not a malfunction in the system, but results from errors made by you on material you had previously seemed to master. Don't worry: that is the way the system works. In particular, it is not unusual to have a "bad" assessment, one that, for external reasons (bad mood, distractions, etc.), does not reflect your actual knowledge. **ALEKS** will quickly bring you back to where you belong.

Problems Too Difficult

It is important to keep in mind that **ALEKS** will not offer concepts that it considers you already to have mastered. Rather, it presents material that you are currently most ready to learn. When the system gives problems that are too hard, the reason is often that you received help or guidance during the assessment or in the Learning Mode. This situation will soon correct itself if you have difficulty with the proposed concepts. The system will revise its estimate of your knowledge and offer concepts that you are more ready to learn.

Repeated Final Assessments

You may need to take more than one final assessment even after you have filled in your pie (in the Learning Mode). This is normal, since mastery is determined by the assessment, not by the Learning Mode. The system needs to confirm (in the assessment) that the entire curriculum has been mastered.

Printing Problems

To print **ALEKS** output (for instance, an Assessment Report) you must press the **ALEKS** "Print" button (on the **ALEKS** menu bar). This opens a new browser window containing the contents of the previous window in the form of a "Print Preview." When this page has been printed it should be closed to return to the normal **ALEKS** interface.

Index

Access Code 2
America Online requirements 2
Answer Editor, help with 9
Answer Editor, purpose of 5
Assessment Report, viewing 9
assessments, final 19
assessments, lengthy 18
assessments, purpose of 5
assessments, results of 7
assessments, rules for 11
assessments, scheduling of 5
buttons, sticky 17
calculators, use of with ALEKS 10
Course Code 2
crashing, how to fix 18
Dictionary, searching 9
FAQ 11
features in ALEKS 8
freezing, how to fix 18
frequently asked questions 11
guidelines for ALEKS use 10
help, online 9
installation of ALEKS plugin 10
Internet access 2
Learning Mode, access to 8
Learning Mode, progress in 8
Learning Mode, rules for 11
logging on to ALEKS 9
Login Name 4
login, unsuccessful 17
Macintosh requirements 2
materials, supplementary 10
messages, how students receive 9
mixed numbers, problems with 18
MyPie 9
Netscape Communicator, incompatible versions 2
Password, changing 9
Password, obtaining 4
PC requirements 1
piechart, interpretation of 7
piechart, reduced 18
plugin, downloading and installing 10
printing, problems 19
problems, too difficult 19
registration in ALEKS 2
regularity of ALEKS use 11
reviewing past material 9
slowness, how to fix 18
technical requirements 1
troubleshooting 17
Tutorial, purpose of 5
typing input, problems 18
User's Guide 1